SPOTLIGHT ON NATURE

IGUANA

LISA M. BOLT SIMONS

CREATIVE EDUCATION · CREATIVE PAPERBACKS

Published by Creative Education and Creative Paperbacks
P.O. Box 227, Mankato, Minnesota 56002
Creative Education and Creative Paperbacks are imprints
of The Creative Company
www.thecreativecompany.us

Design and production by Blue Design, Inc.
Art direction by Tom Morgan
Edited by Jill Kalz

Photographs by Dreamstime/Brian Lasenby, 6, Matildaread, 29; flickr/Biodiversity Heritage Library, 12; Getty Images/Aleksandr Zubkov, 3, 6, 28–29, Antonio Busiello, 24, Anushri Konher / 500px, 15, DAJ, 8, 10, 14, 16, 20, 22, Wild Horizon, 16, Zen Rial, 10; Pexels/Pixabay, cover, 1; Shutterstock/NagyDodo, 18, rebvt, 14, Vikulin, 21; SuperStock/Tui De Roy/Minden Pictures, 29, ZSSD/Minden Pictures/SuperStock, 21; Unsplash/Alex Perez, 11, Denny Luan, 17, Kotagauni Srinivas, 4–5, Mohammad Mardani, 27, Muh Mulyadi, 23; Wikimedia Commons/Benjamint444, 9, Michael Howard, 28

Library of Congress Cataloging-in-Publication Data
Names: Simons, Lisa M. Bolt, 1969- author.
Title: Iguana / by Lisa M. Bolt Simons.
Description: Mankato, Minnesota : Creative Education, Creative Paperbacks, [2026] | Series: Spotlight on nature | Includes bibliographical references and index. | Audience: Ages 10-13 | Audience: Grades 4-6 | Summary: "Strut into the iguana's world with this nature title for middle-grade wildlife lovers. Informational text pairs with a narrative about a single lizard family to spotlight the wild animal's life cycle, supported by infographics and a timeline of developmental milestones"—Provided by publisher.
Identifiers: LCCN 2024048409 (print) | LCCN 2024048410 (ebook) | ISBN 9798889896234 (library binding) | ISBN 9781682777893 (paperback) | ISBN 9798889897033 (ebook)
Subjects: LCSH: Iguanas—Juvenile literature.
Classification: LCC QL666.L25 S56 2026 (print) | LCC QL666.L25 (ebook) | DDC 597.95/42—dc23/eng/20241223
LC record available at https://lccn.loc.gov/2024048409
LC ebook record available at https://lccn.loc.gov/2024048410

Printed in India

CONTENTS

MEET THE FAMILY 4

Green Iguana of Palo Verde National Park

LIFE BEGINS 7

FEATURED FAMILY

Welcome to the World 8

First Meal 10

EARLY ADVENTURES 13

FEATURED FAMILY

Seeking Protein 14

Give It a Try 16

LIFE LESSONS 19

FEATURED FAMILY

This Is How It's Done 20

Practice Makes Perfect 22

THE IGUANA BALANCE 25

Family Album Snapshots 28

Words to Know 30

Learn More 31

Index 32

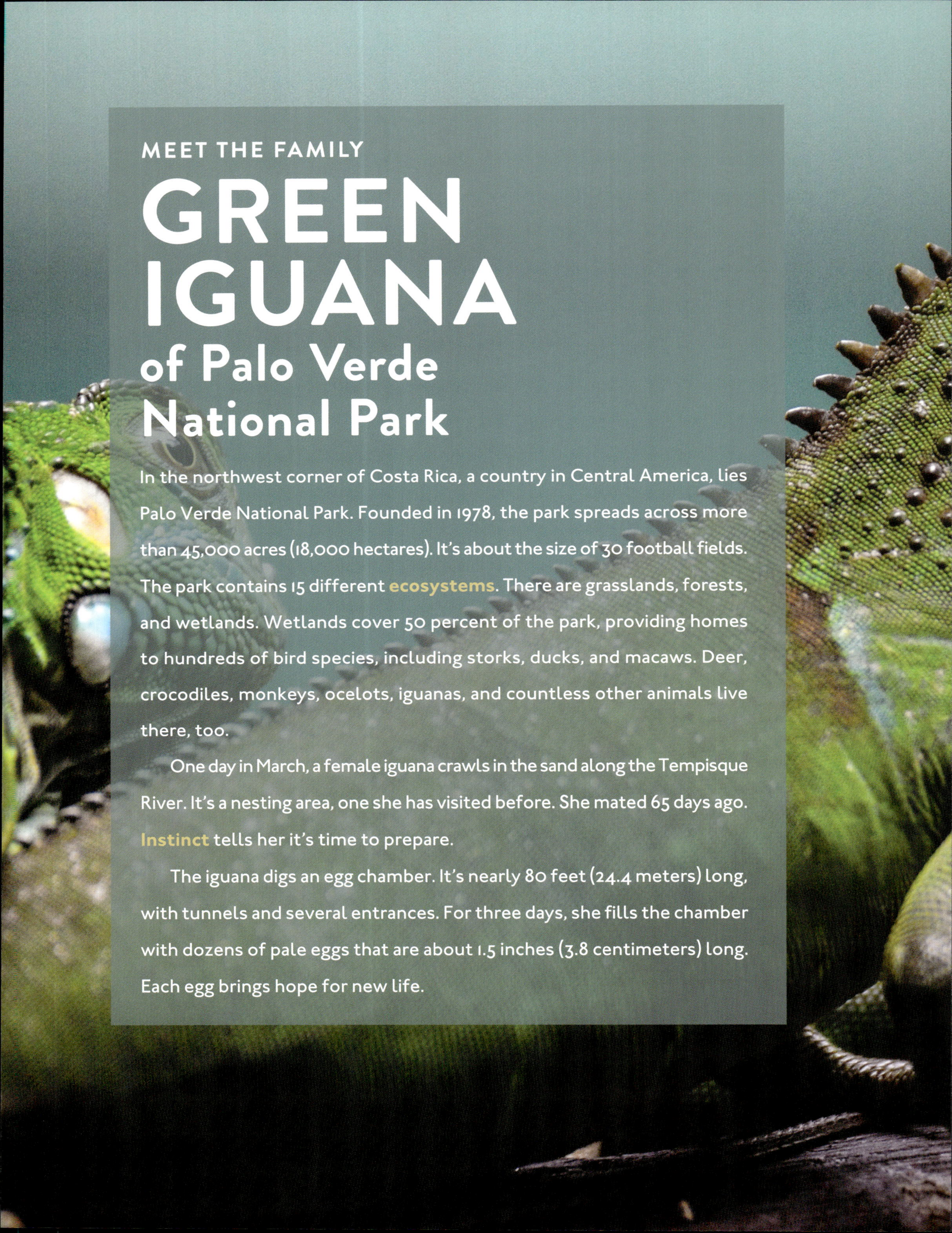

MEET THE FAMILY

GREEN IGUANA of Palo Verde National Park

In the northwest corner of Costa Rica, a country in Central America, lies Palo Verde National Park. Founded in 1978, the park spreads across more than 45,000 acres (18,000 hectares). It's about the size of 30 football fields. The park contains 15 different **ecosystems**. There are grasslands, forests, and wetlands. Wetlands cover 50 percent of the park, providing homes to hundreds of bird species, including storks, ducks, and macaws. Deer, crocodiles, monkeys, ocelots, iguanas, and countless other animals live there, too.

One day in March, a female iguana crawls in the sand along the Tempisque River. It's a nesting area, one she has visited before. She mated 65 days ago. **Instinct** tells her it's time to prepare.

The iguana digs an egg chamber. It's nearly 80 feet (24.4 meters) long, with tunnels and several entrances. For three days, she fills the chamber with dozens of pale eggs that are about 1.5 inches (3.8 centimeters) long. Each egg brings hope for new life.

CLOSE-UP

Caruncle

Baby iguanas have a special "egg tooth" on the tip of their snout. It's called a caruncle. It helps the hatchling break out of its shell. After the young iguana hatches, the caruncle falls off.

CHAPTER ONE

LIFE BEGINS

Iguanas are diverse reptiles. Scientists recognize 35 species in 8 groups. The species range from the spinytail iguana, at 5 inches (12.7 cm) long, to the 7-foot (2.1-m) green iguana. The heaviest species is the blue iguana. It can weigh up to 30 pounds (13.6 kilograms).

Iguanas look a bit like mini dinosaurs. They can be almost black, bright orange, blue and green, green, or green with pink. The color of their scaly skin depends on the species as well as the temperature of the environment and the animals' health and mood. Iguanas have sharp claws. They also have sharp teeth and a thick, slightly forked tongue. Their eyes are usually brown. Instead of external ears, inner ears are covered with a special scale, one on each side of the head. Iguanas have a long tail measuring about half the length of their body. Most species have spines on their tail. These features warn predators to stay away.

Iguanas live alone or in groups. The largest males lead groups made

GREEN IGUANA MILESTONES

DAY 1

- Breaks out of shell
- Digs out of burrow
- Is without parents
- Starts feeding on absorbed yolk

1 WEEK

- Continues feeding on absorbed yolk

up of males and females. Iguanas fight mostly over the best sunny spots. Males are more hostile during breeding season, when they fight over females and territories.

Iguanas live mostly in Mexico, the countries of Central America, and Brazil. They also live on many islands in the Caribbean Sea and Pacific Oceans. They are considered **invasive** in U.S. states such as Florida and islands such as Puerto Rico. Some iguanas live in trees in tropical rainforests. Other species live in the desert or on the slopes of volcanoes.

Iguanas eat mostly leaves, flowers, and fruit. They are considered plant-eaters, or herbivores. Marine iguanas swim in the ocean and eat seaweed and plant-like organisms called algae. Some iguana species also eat small insects and are considered **omnivores**.

Iguanas are amazing swimmers and prefer to live near water. They'll

CLOSE-UP

Tail

Iguanas use their tails for balance while walking or climbing. They also use it like a whip in self-defense. If a predator catches an iguana by the tail, the iguana may detach the tail and run away. Tails usually grow back in weeks to months.

FEATURED FAMILY

Welcome to the World

The female iguana finishes laying her eggs in the maze-like burrow. She has laid about 35. Using her feet, she covers the eggs with sand and then crawls away. She might return to the nest someday, but she never guards it. For the next 100 days, the sun warms the burrow and incubates the eggs. Inside the eggs, great changes are happening as the young iguanas grow and develop. Then one day, at about the same time, the hatchlings break out of their shells and take their first breath.

often escape from predators by jumping into water. But when they first hatch, the baby lizards, left alone by their mother, are easy prey. Hatchlings must fend for themselves, so they stay in and around the burrow at first. However, after about a week, they must try to find food to start to grow. Hatchlings may stay in small groups to better see snakes, mongooses, hawks, and other predators.

(2) WEEKS

- Starts feeding on insects, spiders, eggs, and snails

(4) WEEKS

- Starts to shed skin every 4 to 6 weeks
- Sheds in small pieces instead of one large piece

CLOSE-UP
Dewlap

Mature male iguanas have a throat fan, or dewlap. It warns predators to stay away by making the iguana look bigger than it really is. The dewlap also helps controls the lizard's body temperature.

FEATURED FAMILY

First Meal

One hatchling, a bright-green, tiny version of its mother with no spikes down its spine, climbs out of its eggshell. For the first week or two, it doesn't eat. It lives off the yolk that was absorbed in its body when it was inside the egg. But then the hatchling needs to find food. It finds safety in numbers by joining its burrow mates. They help protect each other from predators as they search the vegetation near the river. Because the hatchling needs protein to grow, it looks for insects and eggs to eat.

IGUANAS ARE EXCELLENT SWIMMERS.

① YEAR

- Starts to shed once a year

② YEARS

- Has learned to swim

CLOSE-UP

Teeth

Even though most iguanas aren't meat-eaters, their teeth are very sharp. The teeth help cut food into small pieces that can be swallowed. Iguanas also use their teeth to inflict painful bites when attacked.

CHAPTER TWO

EARLY ADVENTURES

Baby iguanas stay with other newly hatched iguanas. There is relative safety in numbers. But as they leave the protection of the burrow, hatchlings face many dangers. Without parents to keep them safe, they may fall prey to animals such as hawks and snakes. Wild cats and dogs are threats, too. Despite the dangers, hatchlings must still venture out and find food.

Marine iguana hatchlings, native to the Galápagos Islands, must first learn to swim to find food. With flattened tails and arms to the side, they wave through the water, wary of attacks from hawks, owls, and herons. Chuckwallas, part of the iguana family, live in rocky deserts. Hatchlings learn to hide in small crevices from predators such as coyotes. They take a deep breath, puff out their lungs, and wedge themselves in place.

Even though iguanas have sharp teeth, they don't chew their food. Instead, they use their tongue to handle it and only bite pieces small enough to

3 YEARS

- Is five times its birth weight

CLOSE-UP

Claws

The green iguana lives in trees. It has sharp claws at the end of long toes to help it grip tree bark. Other iguana species have claws to grasp rock.

FEATURED FAMILY

Seeking Protein

The hatchling's small group grows smaller. The baby iguanas go in different directions as they search for food. Some, including the hatchling, have taken to the trees. They are well adapted to climbing them, thanks to their developing claws and tails. Others have been snatched by predators. The hatchling's tongue darts in and out, searching for food. It finds a bird's nest of eggs. Using its tongue, the hatchling moves an egg closer. With its sharp teeth, it bites into the shell, takes small pieces, and swallows.

swallow. At first, baby iguanas eat mostly insects and eggs. As they grow, they eat mostly plants.

Unlike some other animals, baby iguanas don't change much during their first two or three years of life. Baby birds grow feathers. Caterpillars turn into butterflies. Young iguanas just grow longer—fast. Eventually, their tails are at least half their body length. But to live that long, hatchlings have to stay out of harm's way.

3 YEARS

- Most iguanas reach reproductive maturity

CLOSE-UP

Skin

Iguana hatchlings shed their skin every 4 to 6 weeks. Adults shed about once a year. Even the skin on an iguana's spikes is shed. The process of shedding old skin is called molting.

FEATURED FAMILY

Give It a Try

It's midday. The hot sun beats down on the tree canopy. The temperature is 88 degrees Fahrenheit (31 degrees Celsius). A pair of small wild cats called jaguarundis hunt quietly in the cooler shadows far below. The young iguana flicks its tongue. It senses protein nearby. The prey is a large golden silk orb-weaver on the edge of its web. The young iguana tries to catch it with its tongue, but the spider is too fast. It scurries along its web and is gone.

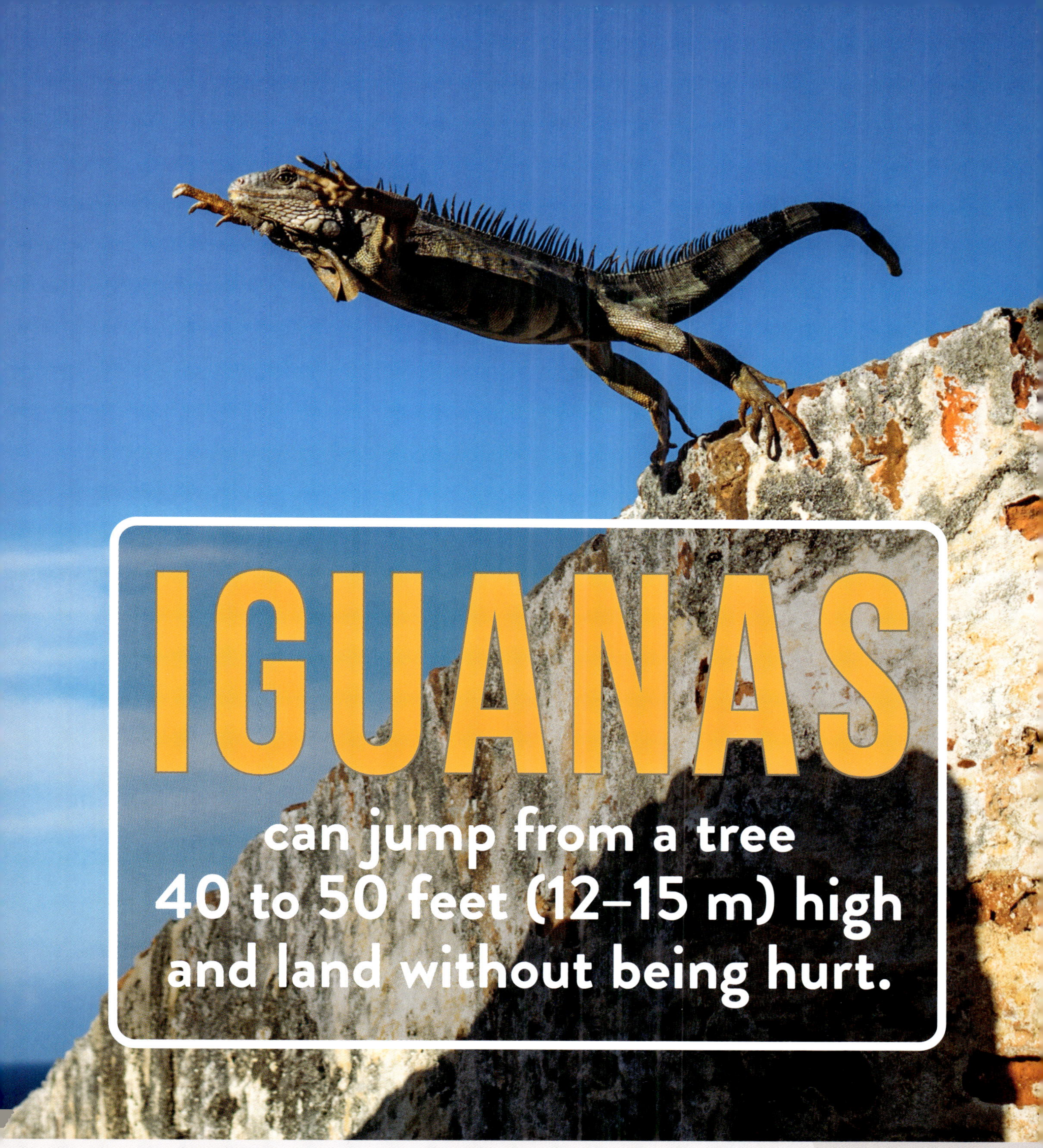

IGUANAS

can jump from a tree 40 to 50 feet (12–15 m) high and land without being hurt.

3 YEARS

- Females lay first eggs
- Deposits eggs in three days

CLOSE-UP

Parietal eye

Behind an iguana's eyes is a sense organ called the parietal eye. This "third eye" helps monitor how much sun the animal is getting. It senses light and dark and also helps the iguana "see" predator shadows.

CHAPTER THREE

LIFE LESSONS

As hatchlings grow, they quickly adapt to their environment. Cuban iguanas learn to share trees with monkeys. Marine iguanas swim with sea turtles. Desert iguanas crawl over sand with tortoises. Different iguanas have varied lifespans and threats to their species. Cuban iguanas can live to be 15 years old, but habitat loss and predators such as cats, dogs, and pigs may shorten their lives. Marine iguanas, the only lizard to spend time in the ocean, can live up to 12 years. Their main threats are dogs, cats, oil spills, and plastic pollution. Desert iguanas can live to be seven years old. Habitat loss and people capturing this species for pets are main threats. No matter where iguanas live, all of them have an instinct to survive. Their ultimate goal is to mate.

At age three, green iguanas are ready to mate. They can mate earlier, but some may not mate until they're five years old. Males court females within a particular territory and mark the females, branches, and rocks with a substance from their body. Males try to get a female's attention by extending and retracting their dewlap. They may also bob their head.

After mating, a female iguana is pregnant for 59 to 84 days. She digs a burrow and lays eggs in it. She does not stay and guard the burrow, nor does she return to it after her young hatch more than 90 days later. The baby iguanas never know their mother. Female iguanas can mate once a year, every year.

Depending on the species, iguanas can live up to 20 years in captivity. In the wild, their lifespan decreases by half.

FEATURED FAMILY

This Is How It's Done

The young iguana has spent its first year eating a lot of insects, eggs, and even some dead animals. It needs the protein for all the growing it's been doing. The reptile snacks on caterpillars and beetles. It munches on a star fruit. Moving from branch to branch, it spots two male iguanas bobbing their heads and showing off their dewlaps. They're fighting over resting sites in the sun. The young iguana watches, learns, and stays out of the way.

CLOSE-UP

Subtympanic shield

What looks like another eye on the head of some iguana species is actually a large scale. There's one on each side of the head, on the jowls. The shields serve no other purpose than to fool predators into thinking that they are getting stared down.

(10) YEARS

- End of life in wild

Iguanas play an important part in their ecosystems. Because they're plant-eaters, they keep vegetation at manageable levels. After they eat fruit, they deposit undigested seeds in their waste onto the soil. There the seeds grow and create new plants. The vegetation not only serves as food and shelter for the iguanas but also for countless other animals. Because of the way they can change their habitat, some land iguanas are called "ecosystem engineers." They support many living things around them. Losing any iguana species to extinction would be tragic.

CLOSE-UP

Spikes

Some iguana species have spikes on their bodies. Green iguanas have a row down the center of the neck, back, and upper tail. These sharp spikes on the tail can help with defense. The common chuckwalla has no spikes.

FEATURED FAMILY

Practice Makes Perfect

The young iguana is almost two years old. It eats during the day when the weather is warm and sunny. Iguanas are **diurnal**. They're also **ectothermic**. Like all reptiles, their activity depends on the temperature. Iguanas cannot control their own body temperature. Once the young iguana has filled its belly with a lot of sweet, juicy fruit called jocotes, it lies in the sun. This rest helps the iguana digest its food.

One of the
LARGEST LIZARDS
in the Americas, the iguana is at once colorful and dull, flashy and plain.

20 YEARS

- End of life in captivity

CHAPTER FOUR

THE IGUANA BALANCE

In 2015, a group of college students and their professor found a fossilized iguana burrow on the island of San Salvador in the Bahamas. It dated back 115,000 years. It was quite a find, showing the long history of rock iguanas in the area. Unfortunately, the species is now endangered on the island.

The story is the same for other iguana species as well. Poachers illegally hunt the animals in large numbers for their meat and eggs. Traders trap iguanas and sell them around the world as pets. Some species, such as the Fiji banded iguana, are at risk of dying out completely.

On the other hand, green iguanas are so numerous that they're considered invasive in the United States. They're threatening the ecosystem in different ways. They destroy people's gardens and eat plants that many butterflies need to reproduce. The decline in insects means less food for insect-eaters. Instead of digging new burrows, green iguanas are using owl burrows, which disrupts owl life cycles.

The iguana species is out of balance. Many organizations are trying to help. The International Iguana Foundation works with a variety of organizations to save iguanas from **extinction**. The foundation also does research, restores habitats, and partners with local communities.

The International Union for Conservation of Nature (IUCN) has an Iguana Specialist Group. Its goal is to help fund research, conservation projects, and "headstarting" to bring back critically endangered iguana species. Headstarting is when baby iguanas are raised in a safe environment before being released into the wild.

The San Diego Zoo built the Griffin Reptile Conservation Center, as well as a new building in Puerto Rico. The zoo helps raise Fiji banded iguanas. It was the first zoo to hatch baby Anegada iguanas, considered critically endangered, and helped with the recovery program of the Grand Cayman blue iguana, too.

Florida is seeing some success with its Exotic Pet Amnesty Program. If iguana owners find themselves unable to care for or no longer want to own their reptile, the pet can be given to the Florida Fish and Wildlife Conservation Commission, no questions asked. New owners with required permits adopt the unwanted iguanas.

Creating a balance for iguanas is key. It will require help from many areas to decrease the number of invasive species and boost those that are endangered.

FAMILY ALBUM

SNAPSHOTS

Because **green iguanas** are raised in Central and South America for food, they are sometimes called "chicken of the tree."

Green iguanas can hold their breath for up to 30 minutes.

Fiji banded iguanas are green with blue bands. They turn darker in the sun to absorb more of the sun's rays.

An adult Fiji banded iguana grows to be six new crayons in length, lined up end to end. The reptile reaches 21 inches (53.3 cm) long.

It's estimated that there are just over 200 **pink land iguanas** left. They are found on Wolf Volcano on Isabela Island off the coast of Ecuador.

The pink land iguana has little coloring in much of its skin. Its blood is visible, making the skin look pink.

Marine iguanas, a species that lives on the Galápagos Islands in Ecuador, can dive more than 65 feet (19.8 m) underwater.

Marine iguanas have a gland that takes ocean salt from their bodies. They sneeze the salt out of their nose.

The **desert iguana** is active during the day, even when temperatures are as hot as 115 °F (46 °C).

Crammed in a rock crevice, the desert iguana can expand its body to twice its size so predators can't pull it out.

Spiny-tailed iguanas are one of the world's fastest lizards. They can reach speeds of up to 21.5 miles (34.6 kilometers) per hour.

Spiny-tailed iguanas spend most of their time on the ground. However, they sleep together in groups high in the trees.

WORDS to Know

diurnal awake during the day

ecosystem a community of organisms and their environment

ectothermic incapable of making one's own body heat

extinction being wiped out of existence

incubate to keep eggs warm so the young inside can hatch

instinct a natural impulse or drive to do something

invasive non-native and usually capable of multiplying quickly, in a bad way

jowl cheek

omnivore an animal that eats plants and other animals

LEARN MORE

Books

Gish, Melissa. *Iguanas*. Mankato, Minn.: Creative Education and Creative Paperbacks, 2024.

Hughes, Catherine D. *Little Kids First Big Book of Reptiles and Amphibians*. Washington, D.C.: National Geographic Partners, 2020.

Somaweera, Ruchira, with Stephanie Warren Drimmer. *The Ultimate Book of Reptiles: Your Guide to the Secret Lives of These Scaly, Slithery, and Spectacular Creatures*. Washington, D.C.: National Geographic, 2023.

Websites

"Fiji Iguana." San Diego Zoo Wildlife Explorers.
https://sdzwildlifeexplorers.org/animals/Fiji-iguana

"Grand Cayman Blue Iguana." Smithsonian's National Zoo & Conservation Biology Institute.
https://nationalzoo.si.edu/animals/grand-cayman-blue-iguana

"Green Iguana." National Geographic.
https://www.nationalgeographic.com/animals/reptiles/facts/green-iguana

Documentaries

Davis, Michael, dir. *Mission Galapagos*. London: BBC, 2018.

Giddings, Al, and David Clark III, dirs. *Galapagos*. Burbank, Calif.: Warner Brothers, 2007.

Maltais, Bryan. *Into the Forest: Reptiles & Amphibians*. Fort Collins, Colo.: Wilderness Media, 2019.

Note: Every effort has been made to ensure that any websites listed above were active at the time of publication. However, because of the nature of the Internet, it is impossible to guarantee that these sites will remain active indefinitely or that their contents will not be altered.

Visit

GALÁPAGOS ISLANDS

Visit these islands off the coast of Ecuador to see the marine iguana.

Ecuador, South America

SAN DIEGO ZOO

Come see the rock iguana of the West Indies, the most endangered group of lizards in the world.

2920 Zoo Drive
San Diego, CA 92101

INDIANAPOLIS ZOO

Grand Cayman blue, Jamaican, and rhinoceros iguanas live at this zoo.

1200 W. Washington St.
Indianapolis, IN 46222

SMITHSONIAN'S NATIONAL ZOO & CONSERVATION BIOLOGY INSTITUTE

Visit the Fiji banded and Grand Cayman blue iguanas.

3001 Connecticut Ave. NW
Washington, DC 20008

INDEX

burrows, 7, 8, 9, 10, 13, 20, 25
claws, 7, 11, 14
climbing, 8, 10, 14
coloring, 7, 10, 23, 28
diet, 8, 10, 12, 15, 16, 22, 25
eggs, 4, 6, 8, 9, 10, 14, 15, 17, 20
feeding, 7, 9, 10, 12, 13, 14, 15, 20, 22
habitats, 4, 8, 13, 19, 22, 25, 26, 28, 29
hatchlings, 6, 8, 9, 10, 13, 14, 15, 16, 19
invasive species, 8, 25, 26
predators, 7, 9, 10, 13, 14, 18, 19, 21, 29
reptiles, 7, 20, 22, 26, 28
scales, 7, 21
shedding, 9, 11, 16
size, 7, 10, 13, 15, 23, 28, 29
tails, 7, 8, 13, 14, 15, 22, 29
teeth, 6, 7, 12, 13, 14
threats, 9, 13, 19, 25, 28
tongues, 7, 13, 14, 16